from me
5 apr '81

ATSF Color Guide to Freight and Passenger Equipment

Lloyd E. Stagner

Published by
Morning Sun Books, Inc.
11 Sussex Court
Edison, N.J. 08820
Library of Congress Catalog Card Number: 95-075996

First Printing
ISBN 1-878887-45-9

Dedication

This book is dedicated to the memory of Jesse T. Smith (1901-1982) who served in the Santa Fe Railway mechanical department from 1916 to 1968. "J.T." as he was better known, started as a machinist apprentice at Clovis in 1916 and retired as Superintendent of Shops at Los Angeles. He was a roundhouse foreman at the age of 21 and served as master mechanic at Amarillo and La Junta for many years. In that capacity he supervised the repair tracks on these divisions that kept the cars pictured in this *ATSF Color Guide to Freight and Passenger Equipment*, functioning smoothly. "J.T." was also a railroad enthusiast and assisted me in writing my first published article, "The Ultimate Development," the story of the Santa Fe 2-10-4 steam locomotives in August 1975 *Trains*. Mr. Smith was one of the best in the mechanical department, knowledgeable, fair, friendly and loyal to his railroad, 24 hours a day, seven days a week in an era when operating department supervisors had no vacations and few days off.

ACKNOWLEDGEMENTS

A large number of photographers contributed to the photographs in this book and I shall attempt to list them, without, I hope any omissions. Don Ball, George Berisso, Bob's Photos, Craig Bossler, W.W. Childers, Steve Evarts, Emery Gulash, Matt Herson, Dave Hickcox, Dan Holbrook, Earl Holloway, C.H. Humphreys, Owen Leander, Dave McKay, J.B. Moore, Russ Porter, Lou Schmitz, Ed Seay, Jr., J.W. Swanberg, Bob Wilt, Paul C. Winters, Chuck Yungkurth of Rail Data Services, and Chuck Zeiler.

Special thanks go to Jay Miller, Art Riordan, John B. Moore and Keith Jordan. Keith furnished much valuable caption information from the resources of the Santa Fe Modelers Organization.

Without Bob Yanosey, of Morning Sun Books there would have not been a book and I again extend my thanks for him paying me for an enjoyable project.

ATSF Color Guide to Freight and Passenger Equipment

Much has been published concerning the motive power of the Atchison, Topeka & Santa Fe Railway System, including Bulletin 75 of the Railway & Locomotive Historical Society in 1949 and more recently the four all color books published by Morning Sun Books, Inc. Certainly, the massive 4-8-4s and WarBonnet PA diesels are worthy of all the attention that has been paid to them. Frequently overlooked are the vehicles that provided the means to earn the revenues, and the cars and machines necessary for the maintenance of the railroad.

Before the "mega-merger" era of the 1970s, the Santa Fe was the most extensive railroad system in the United States. In a 1948 FORTUNE magazine article, Santa Fe was judged the "best railroad" in America, which, of course was controversial. However, the big variety of Santa Fe freight color schemes and advertising slogans painted on the sides of freight equipment surely was not topped. The use of advertising slogans on freight cars had started in 1940 to influence both passengers and freight shippers to use *Santa Fe—All the Way*. Several railroads had one "crack" train they advertised on freight car bill boards, but AT&SF promoted six different trains at various times (SUPER CHIEF, CHIEF, EL CAPITAN and *Grand Canyon Line* with maps and *Ship and Travel* slogan; SCOUT, map only; TEXAS CHIEF and SAN FRANCISCO CHIEF with *Ship and Travel* only). Although the passenger lightweight car paint schemes were unimaginative, with only black lettering, there was a great diversity in the painting of company service cars.

Most Santa Fe freight equipment was painted mineral brown all over until 1930. Reefers had yellow/orange sides and ends with black roofs, and underframes. In 1939-40 reefers began to receive black ends. Maps *and* slogans were used from 1940 to 1947. From 1947 till 1959, the map was eliminated, the slogans were revised and the *Ship and Travel, Santa Fe* slogan also came into use. In 1959 the big circle Santa Fe herald and Indian red paint (for *Shock Control* cars) became the norm. In the mid 1950s all-yellow was used on reefers. In 1966, cabooses blossomed forth in bright red, after almost a century of mineral brown.

Santa Fe classified its revenue freight cars as follows: Bx box car, FE Furniture cars which also included automobile cars, FT Flat car, GA Gondola which included hoppers, covered hoppers, ballast, air dump and other open top cars, LG Log car, RR Refrigerator which included insulated box cars, SK Stock car, IE Ice car, and TK tank car. These designations

are used beneath each freight car photo. Since there were so many class designations in each group, the cars are grouped numerically by each type in this book.

Santa Fe freight traffic was well diversified. For example, in 1965 the following number of carloads were handled of various important commodities: wheat 104,174 carloads, other grain 79,529, grain products 63,834, forest products 105,590, petroleum products 123,754, livestock 40,234, coal and coke 57,396, less-than-carload and forwarder merchandise 146,305 and manufactures, miscellaneous and all other 960,112. Included in all other was a heavy tonnage of fruit and vegetables that moved seasonally from California and Arizona, requiring the ownership of 14,309 refrigerator cars, supervised by S.F.R.D. (Santa Fe Refrigerator Department). During some years, AT&SF was the largest originator of livestock and in 1950 owned 7042 stock cars.

Since the 1895 inauguration of THE CALIFORNIA LIMITED and the subsequent DELUXE in 1911, THE CHIEF in 1926, SUPER CHIEF in 1936, EL CAPITAN in 1938, TEXAS CHIEF in 1948, KANSAS CITY CHIEF in 1950 and ending with the SAN FRANCISCO CHIEF in 1954, Santa Fe was a leader in the operation of fine passenger trains. This facet is covered in the four Santa Fe all-color books published by Morning Sun, and needs no further explanation. Suffice to say, that Santa Fe bought new passenger cars in 1964, when most American railroads were busily seeking ways to discontinue their passenger service.

Railfans and modelers who are interested in further information on Santa Fe equipment are referred to four excellent books published by the Santa Fe Modelers Organization (Now Santa Fe Railway History and Modeling Society). These are "Listing of Santa Fe Freight Cars by Class and Car Numbers 1906-1991," by Larry Occhiello, "Santa Fe Work Equipment Cars" by W.W. Childers "Santa Fe Painting and Lettering Guide for Model Railroaders" by R.H. Hendrickson and "Santa Fe Refrigerator Cars" by C.K. Jordan, R.H. Hendrickson, J.B. Moore and A.D. Hale.

Through the pages of this *ATSF Color Guide to Freight and Passenger Equipment*, we will present, a comprehensive, but not all-inclusive look at the rolling stock that was in service between 1945 and 1980. *Santa Fe All The Way!*

Lloyd E. Stagner
February 1995

LIGHTWEIGHT PASSENGER CARS - "HEAD-END" EQUIPMENT

4

ATSF 3514

▲ Hauling storage mail for Amtrak, 3514 was photographed at Pasadena, CA on October 24, 1973. Thirty of these baggage-mail cars came from Budd in 1953. This series was 73 feet 10 inches long and weighed 94,800 pounds.

(Dave McKay)

ATSF 3421

▼ During the last weeks when U.S. mail was handled by rail, 3421 was the Colorado Springs set-out car on September 21, 1967. It was built by ACF in 1947.

(Dave McKay)

ATSF 3667

▲ 3667 was among the last of the 72 feet 10 inches baggage cars built by ACF in 1955. On October 31, 1971 it was at Chicago in Amtrak service.   *(David H. Hickcox collection)*

ATSF 3534

▼ One of nine baggage-express cars in the 3531-3539 series built by Budd in 1953. 3534 was almost 15 years old when it was photographed on January 20, 1968. After the fall 1967 diversion of U.S. mail from the railroads, many of these cars were idled.   *(David H. Hickcox collection)*

ATSF 3464

The 3464 was an earlier stainless steel baggage car, one of 14 built by ACF in 1950, weighing 45 tons when empty.

(P.C. Winters)

ATSF 3713

The 100 baggage cars built at the Topeka shop between 1956-1958 were 63 feet, 10 inches in length of which 3713 is an example.

(P.C. Winters)

ATSF 80

Railway Post Office 80, was one of two Budd built cars built in 1940 and when delivered assigned to Nos. 11 and 12. These cars were 63 feet two inches in length and weighed 87,650 pounds. *(P.C. Winters)*

ATSF 99

▲ R.P.O. 99 was one of 12 constructed by Budd in 1964 (very late for such traffic), and was out-of-work in the fall of 1967 when all Railway Post Office service was discontinued by the post office department. It was in Kansas City-La Junta service in June 1967. *(Emery Gulash)*

ATSF 140

▼ In 1970, ten steam-generator cars were rebuilt from baggage cars for use mostly with freight F45 diesels working passenger trains in the waning days of Santa Fe service. 140 was the last, converted from baggage car 3931 in July 1970. In 1974 it became 138.

(Owen Leander, Robert J. Yanosey collection)

LIGHTWEIGHT PASSENGER CARS COACHES - CHAIR CARS

ATSF 2798

Chair car 2798 was originally a **SUPER CHIEF** four drawing room, one bedroom, lounge-observation named *Vista Plains*, built by Pullman in 1938. It was subsequently converted for mid-train service (note the rounded end) and named *Coconino*. In 1962, it was rebuilt to a leg rest chair car at Topeka, where it was photographed May 3, 1970.

(Owen Leander, Robert J. Yanosey collection)

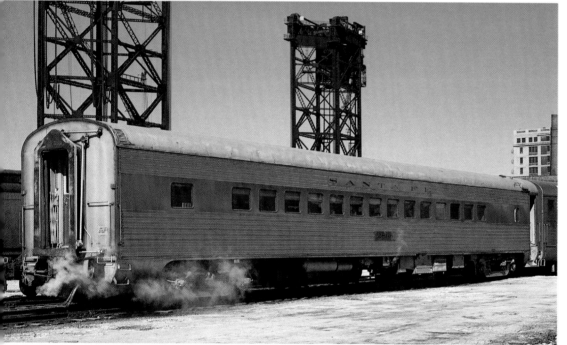

ATSF 2808

Another of the 25 sleeping cars converted to chair cars between 1960-1962, 2808 was built in 1938 as *Tonto*, a 17 roomette car and rebuilt in June 1960. It was at Chicago March 6, 1971, shortly before Amtrak assumed passenger train service.

(Owen Leander, Robert J. Yanosey collection)

ATSF 2845

Forty-five new leg rest chair cars came from Budd in 1953, the last equipment purchased before the change to hi-level equipment. 2845 was in Amtrak service at Chicago on April 14, 1973, part of the 2816-2860 series which accommodated 46 passengers.

(Owen Leander, Robert J. Yanosey collection)

ATSF 2921

▲ 44 seat chair car 2921 was equipped with leg rest chairs when it was built by Pullman in 1950, as one of an order for 33 cars, numbered 2912-2945. On May 3, 1970 it was outside the passenger car shop at Topeka awaiting its last overhaul.

(Owen Leander, Robert J. Yanosey collection)

ATSF 3138

▼ Built as a 60 seat chair car by Budd in 1941, 3138 had been re-equipped with 42 leg rest chairs long before it was photographed at Chicago on March 6, 1971. When delivered 3138 was assigned to the EL CAPITAN, GOLDEN GATE and SAN DIEGAN streamliners.

(Owen Leander, Robert J. Yanosey collection)

9

ATSF 540

Hi-level chair car 540 was one of 24 built by Budd in 1963-1964, the last Santa Fe purchase of new passenger equipment. Delivery of these cars permitted assignment of hi-level chair cars to the SAN FRANCISCO CHIEF. *(David H. Hickcox)*

ATSF 703

703 was a hi-level chair car built by Budd in 1956, one of 35 to re-equip the EL CAPITAN as a hi-level train that summer. At Chicago on October 31, 1971 it had been Amtrak property since May 1. The 700-724 series cars seated 72 passengers.

(Owen Leander,
Robert J. Yanosey collection)

ATSF 726

Twelve hi-level chair cars similar to the 1956 order, were acquired in 1964 that also seated 72 passengers. The 528-537 series had steps to allow access to regular cars on the train and seated 68 passengers. 726 was operated by Amtrak when it was photographed in Chicago on September 9, 1972.

(Owen Leander,
Robert J. Yanosey collection)

LIGHTWEIGHT PASSENGER CARS - SLEEPERS

ATSF *Indian Pony*

▲ The eleven double bedroom cars such as *Indian Pony* shown here at Chicago were the most luxurious sleepers on the Santa Fe and all twelve cars were assigned to the SUPER CHIEF.

(Owen Leander, Robert J. Yanosey collection)

ATSF *Tesque Valley*

▼ A *Valley* series car, *Tesque Valley* was on No. 24 at Joliet on August 12, 1967 and was the only sleeping car on the GRAND CANYON. Inside it contained 6 sections, 6 roomettes and 4 bedrooms.

(Dave McKay)

11

ATSF *Regal Isle*

Regal Isle was one of fifteen ACF two drawing room, four compartment, four double bedroom sleepers built in 1950 for the re-equipped SUPER CHIEF, which was inaugurated January 28, 1951.

(Owen Leander,
Robert J. Yanosey collection)

ATSF *Regal Pass*

The Pullman-manufactured *Regal* series included seventeen cars built in 1948, of which *Regal Pass*, photographed at Houston in August 1969 is an example. There is a difference in height from the top of the windows to the roof on the two series.

(David H. Hickcox collection)

ATSF *Regal Vale*

Right side of *Regal Vale*, one of the 1948 Pullman-built cars at Chicago on August 22, 1971. During the period that through sleeping cars ran through from New York and Los Angeles, the *Regal* series ran on the New York Central TWENTIETH CENTURY LIMITED and Pennsylvania BROADWAY LIMITED east of Chicago. This service was discontinued in January 1958.

(David H. Hickcox collection)

ATSF *Blue Water*

▲ Sleeper *Blue Water* was at Chicago on June 27, 1971 while in Amtrak service. It was one of 19 ten roomette, three double bedroom, two compartment cars, built by Pullman in 1947 when passenger car production was resumed after the World War II hiatus. *(Owen Leander, Robert J. Yanosey collection)*

ATSF *Pine Gorge*

▼ Delivery in 1950 was made by the Budd Co. of 27 ten roomette six double bedroom stainless steel sleepers in the *Pine* series. *Pine Gorge* was waiting to be shopped at Topeka on May 3, 1970. These sleepers weighed 71 tons when ready for service. *(Owen Leander, Robert J. Yanosey collection)*

13

ATSF *Moencopi*

▲ The sleeper *Moencopi* was built by Pullman in 1938 for service on the CHIEF which was streamlined in February of that year. The four double bedroom, four compartment, two drawing room sleeper was at Chicago on March 24, 1968. The use of Indian names on passenger equipment resulted in much misspelling by conductors, consist clerks and mechanical workers. *(Owen Leander, Robert J. Yanosey collection)*

ATSF *Navajo*

▼ ATSF observation-lounge *Navajo* was built by Budd in 1937 for the first streamlined SUPER CHIEF. With a configuration of 3 compartments, 2 drawing rooms, a bed room and a 14 seat lounge, it was retired in 1957 and subsequently sold to the Intermountain Chapter, NRHS. It was photographed on Sept. 28, 1968 at Lincoln, NE while on a private car trip on the Burlington. *(Lou Schmitz)*

14

LIGHTWEIGHT PASSENGER CARS - DINERS - LOUNGES

ATSF 500

▲ The opulence of travel on the Santa Fe was best experienced while riding in one of six dome-lounge cars in the 500-505 series on the SUPER CHIEF. Built by Pullman in 1950 for the re-equipped SUPER CHIEF, they were designated in the public timetable as a *Turqouise Room-Pleasure Dome Lounge Car*. The Turquoise Room provided a private dining room for special parties who were served from the dining car. The 500 was at Chicago on September 9, 1972.

(Owen Leander, Robert J. Yanosey collection)

ATSF 1349

▼ Club-lounge 1349 came from Pullman in 1947 as part of the first post war delivery of new passenger equipment. These cars could seat 50 passengers and were sometimes used for overflow coach passengers on crowded trains.

(Owen Leander, David H. Hickcox collection)

ATSF *Acoma*

Acoma, number 1370 was at Joliet, IL on August 12, 1967. It was one of seven club-lounge-dormitory cars built by Budd in 1937 for the first streamlined SUPER CHIEF. *(Dave McKay)*

ATSF *Concho*

Club-lounge-dormitory 1378 *Concho* was a single car order delivered by Pullman in 1940. In 1955 it was regularly assigned to the CHIEF Nos. 19-20. It was at Dallas in December 1970 on a special train.

(Rail Data Service)

ATSF 1391

No. 1391 was not named. It was built by Budd in 1946, club-lounge-dormitory and was at the Chicago Terminal coach yard on June 27, 1970. *(Owen Leander, Robert J. Yanosey collection)*

ATSF 1489
36-seat dining car 1489 was photographed at Chicago June 27, 1970. It was one of seven diners built by Budd in early 1942, just before passenger car construction ceased for the duration of World War II.

(Owen Leander, Robert J. Yanosey collection)

ATSF 1564
Santa Fe owned 22 lightweight lunch-counter dining cars, also nine lunch-counter diner dormitory cars. For many years they operated on GRAND CANYON trains 23-24, 123-124. No. 1564 was one of the 16 lunch counter diners, built by Budd in the 1550-1565 series in 1948. They could serve 34 people at the counter and at several tables.

(Owen Leander, Robert J. Yanosey collection)

ATSF 1568
Lunch-counter diner 1568 was one of three Pullman built cars of 1950. At 82 feet 11 inches in length they were seven inches shorter than the Budd cars, but had places for 33 persons. 1568 was running on Nos. 23-24 between Chicago-Winslow when it was photographed at Chicago on March 26, 1967.

(Owen Leander, Robert J. Yanosey collection)

ATSF 508

 ▲ Arrival of eight ''big dome'' lounge cars in 1954, numbers 506-513 built by Budd in 1954 allowed their assignment to EL CAPITAN. When hi-level cars were purchased for EL CAPITAN in 1956, the domes went to the CHIEF. 508 was the third of the series when it was photographed in Chicago.

(P.C. Winters)

ATSF 550

▼ The 550-555 series were lounge-dormitory-dome cars also built by Budd in 1956 for assignment on the Chicago-Oakland SAN FRANCISCO CHIEF. Seat capacity on these cars was 82 instead of the 103 carried on the 506-513 series.

(P.C. Winters)

HEAVYWEIGHT PASSENGER "HEAD END" EQUIPMENT

ATSF 78

▲ Railway Post Office 78, a 60 ft. 9 inch car built by Pullman in 1927 was nearing the end of its life at Los Angeles in May 1966. In November 1967 all R.P.O. service would be discontinued by the Post Office. 78 was converted to a company service tool car in April 1990.

(Alan Miller, Matt Herson collection)

ATSF 369

▼ Horse-Express car 369 was used primarily to transport race horses to and from the tracks at Santa Anita and Del Mar, CA. There would be few more calls for its services on April 6, 1969 as it lay at Chicago. In March 1963 369 had been converted from chair car 3052 built by Pullman chair car in 1929. When it was converted to head end duties, it received roller bearing trucks and grey paint.   *(Dave McKay)*

19

ATSF 1997

▲ A newer horse-express car, 1997 was in storage at Chicago October 22, 1973. It was one of ten, series 1990-1999 built by Pullman in 1930. One of the first all-steel horse express cars, roller bearing trucks were subsequently applied in the early 1950s and it was painted grey to distinguish it from the green (non roller bearing) cars.

(Owen Leander, Robert J. Yanosey collection)

ATSF 1999

▼ The end door of horse-express car 1999 is evident in this photo at Los Angeles in July 1973. By this time, most race horses were being transported by air, and 1999 was waiting a one way trip to the scrapper. *(Bob Trennert collection)*

ATSF 2415

Many mixed trains were operated on Santa Fe branch lines before the wholesale discontinuance of passenger service. 2415 was built as wooden coach 697 in 1905 and converted to a combination-coach-baggage car in October 1932. It was on the Henrietta-St. Joseph, MO Branch, laying over at the latter terminal on February 15, 1958. Generally, heavyweight combines used in branchline/mixed train service were painted in a dark red/brown color. *(Lou Schmitz)*

ATSF 2544

Coach-baggage-express 2544 was painted "coach" green which indicated it was sometimes assigned to main line passenger trains. On heavyweight equipment, Santa Fe generally utilized this "coach" green which was lighter than Pullman green. Another product of 1927 by Pullman Inc. it was at Chicago on September 28, 1969 and donated to the Illinois Railway Museum that same month.

(Owen Leander, Robert J. Yanosey collection)

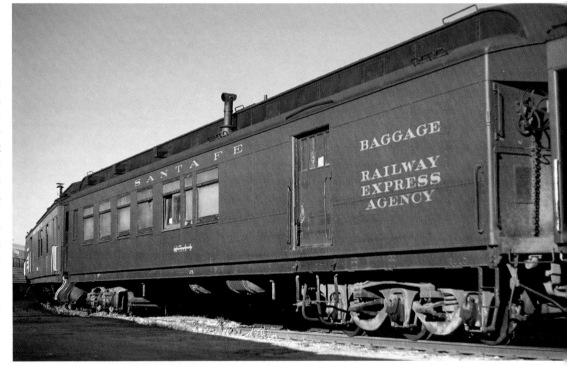

ATSF 2603

The 2602-2608 series of coach-baggage-express cars were usually assigned as "rider" cars for the crews on Nos. 7-8, FAST MAIL EXPRESS. Nos. 7-8 had been discontinued in October 1967 so there was no future for the 2603 when it was at Chicago on Feb. 22, 1968.

(Owen Leander, Robert J. Yanosey collection)

ATSF 2638

▲ 2638 was still needed when mixed train 67 stopped at the Wichita Union Station to load some express on May 7, 1966. Ahead was an eighty mile trip to Pratt, KS and a return to Wichita in the afternoon. A GP7 was the power on 67-68.

(J.F. Porter, J.B. Moore collection)

ATSF 2640

▼ The 2640, was a 1917 Pullman-built coach-baggage-express awaiting disposition at Chicago Sept. 28, 1969. AT&SF ended passenger service on most branch lines in 1968 simultaneously with the reductions in main line service that saw such famous trains as the CHIEF making its last run.

(Owen Leander, Robert J. Yanosey)

ATSF 4342

▲ Although classified as freight cars, 50 foot passenger box cars 4100-4399, built by Pullman in 1941-1942 were primarily used on passenger trains to handle mail and express. All 300 cars were equipped with signal and steam lines and had brackets for markers if moved on the rear of a train.

(F. Waddington, R. Trennert collection)

ATSF 4213

▼ ATSF 4342 *(above)* was on home rails at Barstow, CA on December 2, 1962, while 4213 was photographed at St. Louis since these cars moved interline on some express moves.

(P.C. Winters)

ATSF 208

▲ Long before containers were used for export-import freight, Santa Fe cut down 30 year old baggage cars to flat cars, added roller bearing trucks and purchased several hundred containers to handle mail, primarily on Trains 1-2, SAN FRANCISCO CHIEF, between Chicago-Oakland, CA. Flat car 208 was at Chicago in July 1963.   (P.C. Winters)

24

REX 4022

▼ Railway Express operated 100 expres-refrigerator cars in the 4000-4099 series which were owned by AT&SF Railway. In 1967 the Express Company paid Santa Fe six cents a mile for their use. These 56 feet, one inch cars were assigned principally to perishable express lading. In 1967, only three other railroads, Great Northern, Atlantic Coast Line and Seaboard Air Line leased express refrigerators to the express company. The 4022 was photographed in Chicago on April 6, 1969.   (Dave McKay)

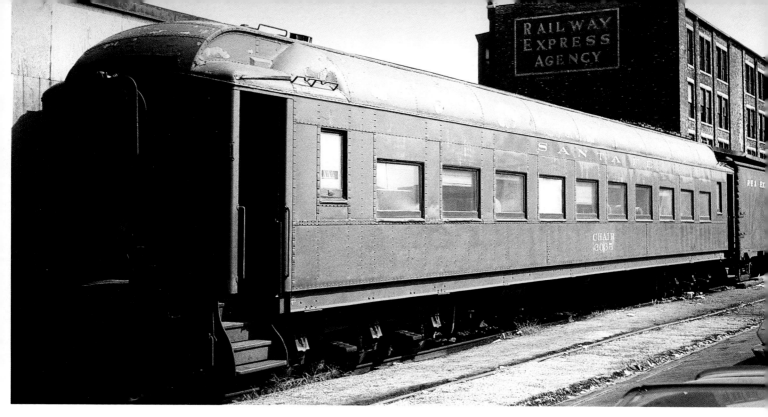

HEAVYWEIGHT PASSENGER EQUIPMENT - CHAIR CARS

ATSF 3035

The sixty-eight 3000-series chair cars were the backbone of main line service until the post World War II delivery of new stainless steel cars. 3035, built by Pullman in 1928 was still in service at Chicago August 20, 1966. *(C.H. Zeiler)*

ATSF 3041

3041 had been equipped with air-conditioning and refitted with 53 reclining seats in the mid-1930s. It left Pullman in 1928 and was still available forty years later at Chicago Feb. 22, 1968 for extra service. *(Owen Leander, Robert J. Yanosey collection)*

ATSF 3067

The last of the 3000-series chair cars, 3067 was built by Pullman in 1930. It was a rebuild with lounge-smoking rooms for both sexes and seated 53 passengers. With an overall length of 70 feet, it weighed 90 tons, compared to 50 tons for the first lightweight chair car, 3070. *(Alan Miller, Matt Herson collection)*

ATSF 1368
78-foot lounge-dormitory 1368 built by Pullman in 1926 was subsequently fitted with roller bearing journals. At Chicago on March 24, 1968 it had run its last mile and was waiting sale for scrap.　　*(Owen Leander, Robert J. Yanosey collection)*

ATSF 1525
Another version of a lounge-dormitory car, 1525 came from Pullman in 1930 and subsequently was fitted with roller bearing journals. Seating only 24 passengers, it was stored at Chicago March 24, 1968 awaiting disposition.

(Owen Leander, Robert J. Yanosey collection)

ATSF 1468
The 1468 was a 46 seat diner that had been retired for several years when it was photographed at Emporia with weathered paint on May 4, 1970. It seated 36 passengers and weighed 98 tons ready for service. Lightweight diners with the same seating capacity weighed only 70 tons. Series 1464-1473 was built by Pullman in 1925-1926.

*(Owen Leander,
Robert J. Yanosey collection)*

ATSF 1303

▲ Baggage-lounge-dormitory 1303 had a 28 feet 3 inches baggage compartment at one end and seats for 34 passengers. In May 1966, it was still in service at Los Angeles, available for service on special trains. Built by Pullman in 1927 as bar-lounge-dormitory 1351 it was rebuilt in 1948 to its later configuration. *(Alan Miller, Matt Herson collection)*

ATSF 1527

▼ The 1527 was a Pullman product of 1930, designated as a lounge-dormitory car, with berths for the dining car crew on transcontinental runs. It was still in service at Los Angeles in May 1966. The two tone grey color scheme of this car and 1303 above is reminiscent of the SCOUT heavyweight scheme that was used from 1936 to 1946.

(Alan Miller, Matt Herson collection)

ATSF 3366

▲ Built as an 80 seat coach in 1926, 3366 was converted to a diner-snack car in 1956 for use on the Del Mar race track specials that ran out of Los Angeles in late August-early September when the horse races were scheduled. It was still seeing service in September 1964.   *(Rail Data Services)*

ATSF 5000

▼ Diesel instruction car 5000 was built as cafe-lounge-observation 1508 in 1926 and converted as a traveling class room for enginemen in December 1947 at which time it was painted in the grey scheme that was used for modernized heavyweight cars between 1947-1970.

(Owen Leander, Robert J. Yanosey collection)

ATSF 5006

▲ Kansas Territorial Centennial Exhibition car 5006 had a varied history. Built as the Pullman sleeper *Madras* in 1916, it was converted to a tourist sleeper by the Pullman Company in 1935, numbered 4158. In 1953, AT&SF bought the car and converted it to company service sleeping car 196464. In January 1954, it was rebuilt for use as an exhibition car and toured the state of Kansas. Later in 1954 it was retired but not sold for scrap until 1970. On May 4 of that year it was laid-up at Emporia.

(Owen Leander, Robert J. Yanosey collection)

ATSF 5005 *Chisholm Trail Centennial Museum*

▼ Built in 1930 as smoker-chair car 2952, renumbered 3122 in 1936, then converted to lounge-dormitory 1527 in 1948, the 5005 assumed that number in 1967 when further rebuilt to display car 5005 to commemorate the centennial of the Chisholm Trail which was used to drive cattle overland from Texas to Kansas railheads. It was painted dark blue and gray for the *Panorama of Progress* display in 1970, and finally received a red, white and blue scheme for the *Spirit of '76* train in 1975. It had been re-numbered 87 in 1973. This photo was taken at Topeka in May 1968.

(Marion Perrin, Lou Schmitz collection)

 ## BUSINESS CARS

ATSF 404

The 400-series "short" business cars were assigned to division superintendents. 404, built by Pullman in 1925, was assigned to the Eastern Division at Emporia, KS on Feb. 8, 1955.

(Don Ball collection)

ATSF 405

Business car 405 was assigned to the Western Division at Dodge City, KS until that division was consolidated with the Middle Division on Sept. 1, 1958.

(Emery Gulash)

ATSF 407

The 407 still retained its original paint at La Junta, CO headquarters of the Colorado Division on August 17, 1965. Although these cars were only 52 feet in length, they weighed 77 tons.

(C.H. Zeiler)

ATSF 37
Heavyweight 37 was built by Pullman in 1925 and was at the Chicago passenger car yard on June 27, 1970.
(Owen Leander, Robert J. Yanosey collection)

ATSF 39
Another business car at Chicago on June 27, 1970 was the 39, also Pullman-built in 1928. As of 1967, AT&SF rostered 15 large and 15 small business cars of heavyweight construction, plus four lightweight cars.
(Owen Leander, Robert J. Yanosey collection)

ATSF *Santa Fe*
Santa Fe was assigned to the President, who on April 7, 1971 was Mr. John S. Reed. It and *Topeka* were delivered by Budd in 1957. *(Owen Leander, Robert J. Yanosey collection)*

ATSF *Atchison*

Atchison was named *Santa Fe* when built in 1949 by Pullman and assigned to President Gurley. In 1957, when the second *Santa Fe* was built, this car became *Atchison* and was assigned to the Vice-President-Traffic.

(C.H. Zeiler)

ATSF 17

Car 17, assigned to the Western Lines Assistant General Manager, was tied-up at Amarillo Aug. 18, 1964. It had been converted from coach 3361, a 1928 Pullman product, in 1952.

(J.W. Swanberg)

ATSF 54

Car 54 at Topeka Dec. 24, 1976 had an interesting history. Built for the United States Railway Administration in 1918 for President Ripley of the Santa Fe it was numbered 31 when purchased by the AT&SF in 1920 and renumbered to 54 in 1973 and was assigned to the General Manager at Topeka.

(G.H. Menge,
Matt Herson collection)

BOX CARS

ATSF 12950 Bx-76

▲ When Santa Fe first began to use "cushion underframes" on its box cars, it was proud enough to adorn the bright red cars with several slogans. *Floating underframe gives freight a smoother ride* was spread across this new car in black when photographed new in 1958 at San Bernardino, CA.

(*K.B. King, Earl Holloway collection*)

ATSF 9256 Bx-76

▼ What a difference twelve years on the road can make! *DF . . . with . . . Shock Control* in white was also used on these XL cars when built in 1958. This more weathered photo was taken in March 1970 at Columbus, OH. Bx-76 was the first Santa Fe class with "Shock Control" (a Santa Fe term). They were painted "Indian red" to distinguish them from the mineral brown, non-shock control cars. (*P.C. Winters*)

ATSF 9870 Fe-29

"FE" was the Santa Fe designation for FurniturE cars long after the original furniture cars had disappeared. Mineral brown 9870 was a XM car built by Santa Fe Topeka shops in 1952 and in May 1974 was at Newark, NJ on the Central RR of New Jersey.   (M.J. Herson)

ATSF 11611 Bx-135

This 50 feet, 6 inches car was built with staggered doors in 1968. Use of the large "billboard" cross and circle herald of nine feet diameter commenced in December 1959.

(E.J. Gulash)

ATSF 6111 Fe-26

Built at the end of World War II in 1945 by Pullman-Standard Car Manufacturing, this staggered double-door 40' box car was photographed at Ardmore, OK on June 15, 1973. In 1973 there were 336 cars of this series on the Santa Fe roster.

(Earl Holloway)

ATSF 11971 Bx-79

▲ A 1961 product of the Topeka shop, 11971 was equipped with a shock-control underframe, "SL" type load restraining equipment and was of the XL mechanical designation. In May 1974 it had been freshly re-painted and the "SL" was advertised in an orange circle in the upper right hand corner.

(Ed Seay Jr.)

ATSF 12021 Bx-79

▼ Santa Fe was nationally known for its bright colored, clean freight equipment and this Bx-79 photographed at Ardmore, OK on March 30, 1977 lives up to the reputation. Notice the absence of the circle "SL." The car was repainted in May 1976.

(Earl Holloway)

ATSF 12873 Bx-75

The TEXAS CHIEF train slogan first appeared in 1948 when that Chicago-Galveston streamliner was inaugurated. At Berea, Ohio on Conrail on May 9, 1976 it still advertised a train that had passed into history five years earlier. 12873 was built by Pullman in 1958 and was equipped with "DF" (damage free) load restraining equipment.

(Dave McKay)

ATSF 14064 Bx-108

This XL type mini hi-cube box car had a high roof for home appliances. Built by Transco in 1967 it was of 40 foot, 6 inches inside length and equipped with "DF" equipment. The photo was taken at Columbus, OH on Penn Central in April 1968.

(P.C. Winters)

ATSF 14597 Bx-81

In fresh paint on July 20, 1974 at Cleveland, OH. This 50 foot 6 inches box car had been built with shock control underframe and "DF" equipment by the Santa Fe in 1960.

(Dave McKay)

ATSF 17106 Bx-69

▲ The use of various passenger train names on freight cars began in 1940 and was still being done in a different version in 1955 when ATSF 17106 left the ACF plant. AT&SF was no longer running the SUPER CHIEF when this photo was taken in 1973. *(Steve Evarts)*

ATSF 17298 Bx-69

▼ Another train name used commencing in 1940 was EL CAPITAN, the Chicago-Los Angeles, all-coach streamliner. The version pictured here was used between 1947-1958. *Ship and Travel Santa Fe All the Way* was painted on the opposite side of the car. *(Steve Evarts)*

ATSF 20673 Bx-85

▲ Bx-85 class cars were rebuilt in 1963 from Bx-12 class cars constructed in 1930, and were among the last 40 foot box car rebuilds. 20673 was at Columbus, OH in February 1964.

(P.C. Winters)

ATSF 22257 Bx-115

▼ Originally built by Pullman in 1942, an effort was made in 1966 to rebuild box cars with plug doors and an orange access door for the grain spout to reduce loss of lading. 22257 was in Chihuahua, Mexico in January 1974.   *(M.J. Herson)*

ATSF 22911 Bx-86

▲ Another example of a 1930 Bx-12 box car totally rebuilt for grain loading, 22911 was fitted with corrugated side walls in 1964 at company shops. Essentially the underframe was the only survivor in this rebuilding. In August 1967 it was at Milwaukee, WI. *(Rail Data Services)*

ATSF 22921 Bx-86

▼ The billboard *Santa Fe* was introduced in 1972 and 22921 still looked freshly painted with an all-yellow plug door at Ardmore, OK May 30, 1973. A 1930 Bx-12 car, rebuilt in 1964 with a re-inforced underframe and new doors, it was assigned for the loading of pet foods at Oklahoma City and retired in the 1970s. *(E. Holloway collection)*

ATSF 22926 Bx-86

By November 1976, this special purpose XM box car had the white XF "Food Loading" sign affixed when photographed at St. Charles, IL. *(D.P. Holbrook)*

ATSF 39385 Bx-126

The *Ship and Travel Santa Fe All the Way* stencilling started in 1959. One of 500 former Bx-53 class cars rebuilt in 1967, 39385 was a standard XM 40 foot 6 inches car with 6 foot doors photographed in February 1976.

(Bill Phillips, M.J. Herson collection)

ATSF 32385 Bx-58

A real rarity at Los Angeles in August 1968, 32385 was originally built by Pullman in 1923 as a wood-sheathed FE-P automobile/furniture car and had been rebuilt in 1950. Nineteen of the three hundred 32000 series cars were still extant in September 1967.

(George Berisso)

ATSF 39872 Bx-136
Rebuilt in 1969, from Bx-53, an original 40 foot car built by AT&SF in 1949, 39872 was equipped with a white painted door signifying its assignment to food lading, in most cases sacked flour. It was at Kent Avenue, Brooklyn in early 1976.

(M.J. Herson)

ATSF 42116 Bx-72
This XM box car had left the Pullman plant in 1957 and apparently had never been repainted when it was photographed at Amarillo, TX in February 1979. Note the black ends and roof. *(A.T. Crittenden, J.B. Moore collection)*

ATSF 42495 Bx-72
The Bx-72 class XM cars, 750 of which were built by Pullman in 1957, had name train slogans typically on the left side and the *Ship and Travel* slogan on the right side. TEXAS CHIEF was on 42495. *(Steve Evarts)*

ATSF 42663 Bx-72
Almost five years after Amtrak took over AT&SF passenger train service, 42663 was at Cleburne, TX January 18, 1976 advertising the CHIEF which made its last trip in May 1968. *(M.J. Herson)*

ATSF 45717 Bx-181
Equipped with *Super Shock-Control* underframe, these cars were popular with shippers of many products including paper, canned and sacked food and household goods. Built in 1976, 45717 was at Reading, PA on Conrail in 1981. *(C.T. Bossler)*

ATSF 43049 Bx-61
Adapted for food loading, 43049 was built in AT&SF shops in 1951. In the fall of 1976 it was at Columbus, OH on Conrail with white doors and "bill board" *Santa Fe*, which was first used beginning in 1972. *(P.C. Winters)*

ATSF 45859 Bx-178
Appliance XM box car equipped with *Super Shock-Control* underframe 45859 was new on July 20, 1975. The 16 foot wide doors facilitated loading and unloading of commodities such as refrigerators and electric stoves. Note the white wrap-around ''excess height car'' band on the car end.
(Ed Seay Jr.)

ATSF 48138 Fe-37
Fe-37 48138 shows off its red paint trimmed with black roof, sill and ends at Columbus, Ohio in May 1967. There are nine colors here with white lettering and logo and a yellow *Super* over the slogan *Shock Control*.
(P.C. Winters)

ATSF 48216 Fe-37
A sister Fe-37, 48216 was almost new in November 1964 at San Francisco, with its Indian Red paint (indicating a *Super Shock Control* car) still fresh.
(Rail Data Service)

ATSF 50058 FE-26
Built by Pullman in 1945, 50058 was one of 299 automobile cars, which were later shifted to general purpose loading as automobile traffic left the rails until the post-1960 development of the automobile rack.

(D.P. Holbrook collection)

ATSF 50256 FE-26
On the Brooklyn Eastern Terminal District Railroad in 1978, 50256 was 33 years old, but had lasted long enough to receive the large Santa Fe emblem and carry it for many years.

(M.J. Herson)

ATSF 128577 Bx-12
3500 50 ton single-sheathed box cars of the Bx-12 class were received in 1930. It would not be until 1936 when additional freight cars were delivered after 1000 more came in 1931, an order the AT&SF would have liked to be able to cancel. 128577 was one of 115 cars that were left in largely original configuration painted mineral brown with black roof when it was photographed in April 1966 while at Columbus, OH on the Pennsylvania Railroad. *(P.C. Winters)*

ATSF 137960 Bx-27

▲ Large purchases of freight equipment were resumed in 1937, of which there were 2500 all-steel XM box cars, including 137960. In 1936, 500 similar Bx-26 box cars were delivered. These cars were the "work horses" of World War II.   *(Rail Data Services)*

ATSF 140613 Bx-34

▼ Bx-34 class comprised 2801 more 50 ton Duryea-cushion underframe box cars delivered in 1940 by Pullman. By 1973, it had acquired the SUPER CHIEF train slogan and would soon be retired. 140613 had the curved system map and name train slogan earlier.   *(Steve Evarts)*

45

ATSF 143585 Bx-37

Built in October 1941 by Pullman, 143585 had acquired the *Grand Canyon Line* slogan when it was built, one of five slogans being authorized for new or rebuilt or repainted cars in January 1940. On March 7, 1973 it was at Ardmore, OK on home rails.

(E. Holloway collection)

ATSF 146112 Bx-37

A wartime product of Pullman built in 1942, 146112 was one of 2000 Bx-37 class 50 ton XM type box cars. The *Ship and Travel Santa Fe All the Way* slogan was first applied to cars in March 1947, replacing the more labor intensive system maps.

(B. Meunier, J.B. Moore collection)

ATSF 146310 Bx-37

This 1942 car had been shopped at Topeka and turned out in April 1973 with the new *Santa Fe* billboard lettering. It was on a siding at Ardmore, OK May 21, 1973. This 40 foot XM car was originally painted in the system map/train name style.

(E. Holloway collection)

ATSF 145388 Bx-33
▲ A 1920 car rebuilt in 1940 as an all-steel box car, 145388 was still on the roster in November 1967. After chair cars or coaches were added to the CHIEF in January 1954, the train slogans were revised to eliminate *All Pullman* before *Chicago-Los Angeles Streamliner*. As built in 1920, this car was of the standard USRA wood-sheathed type.   *(Rail Data Services)*

ATSF 211673 Bx-12
▼ These 1930-built XM 50 ton box cars were built in the company shops at a time when freight traffic was rapidly declining. 21173 was at Columbus, OH when photographed in Dec. 1962.   *(P.C. Winters)*

47

ATSF 212861 Bx-12

▲ This XM car was built as part of the big 1930 order for box cars, but was rebuilt in 1941 with an extended roof for airplane parts. Outside braces made it difficult to paint a slogan on the sides. 212861 was still on the roster in October 1967 with its mineral brown paint in good condition.

(Rail Data Services)

ATSF 213481 Bx-12

▼ This photo reveals some of the details on a Bx-12 50 ton box car. In October 1960 when this was photographed, 213481 was near the end of its career on the Santa Fe. Some Bx-12 cars had wood sheathing replaced with steel in the 1960s.

(Rail Data Services)

ATSF 272198 (2 views) Bx-46
Two excellent photos of ATSF 272198 at San Marcos, TX near the end of World War II in August 1945. On one side, the *El Capitan Coach Streamliner West* slogan appears and on the other (shadowed) side a map of AT&SF main lines and a few branches appears with the slogan *ship Santa Fe all the way*. This painting arrangement was applied to steel sheathed box and automobile cars and all refrigerator cars, both wood and steel, commencing in January 1940 and continuing through April 1947.   *(E.J. Gulash)*

ATSF 274722 Bx-48

One of 750 50 ton XM box cars built by Pullman in 1946, 274722 was still in service at Bensenville, IL on the Milwaukee Road in November 1969. By then, 40-foot box cars with six foot doors were losing the race to piggyback vans and 70 ton, 60-foot box cars.

(Rail Data Services)

ATSF 276778 Bx-151

A rebuild of 1972 from an older Bx-60 class car built by Santa Fe in 1951, the eight foot diameter herald and large *Santa Fe* billboard lettering had been applied to 276778, at Los Mochis, Mexico Feb. 20, 1974. 50-foot cars had nine foot heralds.

(M.J. Herson)

ATSF 501234 Bx-144

The era of high capacity, roller bearing journals, damage free device, cushion underframe box cars began in the 1960s. 501234 in eye catching Indian Red paint was built in 1970, and had just left the Topeka shop after an overhaul when it was photographed at Fort Smith, AR ten years later.

(J.B. Holden, M.J. Herson collection)

ATSF 501670 Bx-169

A 50 foot, 6 inches, 92 ton capacity box car, 501670, built in 1974, was photographed on the Penn Central at Columbus, OH in December 1975. The absence of slogans and any yellow color makes this a "plain Jane" but the Indian Red still stands out against the mineral red Southern car in the background. *(P.C. Winters)*

ATSF 504063 Fe-41

For some reason, AT&SF classified this load divider XL box car in the Fe class. 100 of this series, including 504063 were built in 1969, primarily for canned foods and wine loading in California. *(Ed Seay Jr.)*

ATSF 520976 Bx-114

Santa Fe was still building freight cars in the Topeka shop in 1967 when 700 Bx-114 class cars in the 520750-521449 series were turned out. This colorful Indian red car was photographed in February 1967 handling one of its first loads. *(Bob Wilt)*

ATSF 522373 Bx-146
Pacific Car & Foundry Co. built 300 class Bx-146 70-ton 51-foot XLI box cars for AT&SF in 1971. Typically paint specifications can vary between manufacturers and PC&F used a darker version of Indian Red paint on 522373. *(Rail Data Services)*

ATSF 522765 Bx-167
The PC&F plant at Renton, WA also built 300 more box cars in class Bx-167 in 1975. These were insulated cars with dual-air-packs and load dividers of 52 foot, 6 inches inside length, with 5145 cubic feet and 134000 pounds load capacity.
(Ed Seay Jr.)

ATSF 600194 Bx-88
A 60 foot 8 inches XL box built in Santa Fe shops in 1964, 600194 was off-line at Secaucus, NJ on the former Erie Lackawanna on March 18, 1978. It had been shopped and re-painted since its construction.
(J.C. Smith, M.J. Herson collection)

AUTO PARTS BOX CARS

ATSF 36841 Bx-91

▲ A 1964 product of Thrall Car Co. these 86 foot 6 inches cars with 20 foot all-aluminum doors were in the vanguard of the new high cubic capacity auto parts cars to replace the fleet of 50 foot box cars which had served the automotive industry since the 1930s. 36841 was at Melvindale, MI in late 1964.

(E.J. Gulash)

ATSF 36845 Bx-91

▼ Colorful Bx-91 36845 stands out against a dramatic sky at Melvindale in July 1964. There was no doubt in the observant public's mind as to what was inside the big car as it carried a large white *Auto Parts* stencil which was further explained by "Ford Motor Co." painted next to the Bx-91 class designation.

(E.J. Gulash)

ATSF 36885 Bx-91

▲ Although of the same classification, 36885 had different lettering. The *Super Shock Control* was shifted to the right side and *Auto Parts* eliminated. It had been repainted during a Topeka shopping in January 1976 and was still in pristine condition on the D&H at Colonie, NY on May 28.

(M.J. Herson)

ATSF 36943 Bx-124

▼ Another Thrall Car product of 1967, 36943 was at Alburtis, PA on July 8, 1972, as an example of the weathering of the original Indian Red paint. With a cubic capacity of 10000 feet, and an overall length of 93 foot, 11 inches, these were impressive examples of how the size of freight cars had grown by the 1960s.

(C.T. Bossler)

ATSF 36991 Bx-96

▲ One of an order for only 18 cars, eight-door 36991 was built by Thrall Car Co. in 1964. Weight was not of primary importance in the shipping of many automobile parts and the Bx-96 class had only 80,000 pounds capacity, the standard for box cars built early in the century.   (*P.C. Winters*)

ATSF 37359 FE-36

▼ Although of XAP mechanical designation, 37359 built in Santa Fe shops in 1963 was of 60 foot 9 inches length and 180000 pounds capacity for handling automobile engines and other heavy components. It was photographed at Detroit in October 1963.   (*E.J. Gulash*)

RBL BOX CARS

SFRA 41001 Rr-79

▲ Although of RB (insulated box) mechanical designation, AT&SF designated series 41001-41012 as class Rr-79. These 12 cars were rebuilt at West Wichita car shop in 1964 from box cars which had been previously rebuilt in 1950 from Rr-48 class refrigerators of 1923. Assigned to flour loading out of Wichita, they did not prove particularly useful due to their small size and all were retired by 1972. The square hole above the S initial afforded air ventilation, a feature not available in ordinary XM box cars which were usually employed in sacked flour loading. The "SFRA" reporting marks indicated "in-

sulated box car, no load restraining devices." They were the only "reefers" painted mineral brown (other than ice cars).

(Bob's Photos)

SFRB 5353 Rr-73

▼ One of 350 *Shock Control* insulated RBL cars built in the Santa Fe shop in 1962, class Rr-73 cars were assigned principally to the loading of commodities such as beer or wine that required protection from freezing temperatures. 5353 was at Columbus, OH on the Penn Central in September 1975.

(P.C. Winters)

SFRB 5978 Rr-64

▲ Was there ever a more eye catching paint scheme used on freight cars by an American railroad? 5978 had just left the San Bernardino car shop where refrigerator and RBL type box cars were repaired in February 1974. It was built by Santa Fe in 1959 one of 300 class Rr-64 65 ton capacity cars.

(George Berisso collection)

ATSF 5945 Rr-64

▼ Another Rr-64 SFRB 5945 at Ardmore, OK on May 10, 1973 shows subtle differences between itself and 5978 above. *A smoother ride* is in reefer orange and *Shock Control* is much smaller on this car.

(Earl Holloway)

SFRB 6157 Rʀ-57

▲ The 6000-6249 Rʀ-57 series were the first of the RBL cars, built for Santa Fe in 1955 by Pullman Standard Car Manufacturing Co. The original paint scheme had been modified when it was photographed at Cleveland, OH in March 1972. It was still equipped with friction bearings as the first AT&SF roller bearing freight cars came in 1956.

(Dave McKay)

SFRB 6011 Rʀ-57

▼ The *Ship and Travel Santa Fe* slogan of 6157 was absent on Rʀ-57 6011 photographed at Council Bluffs, Iowa on June 10, 1973. With no more passenger trains to advertise since the Amtrak takeover of 1971, a simple *Santa Fe* would suffice.

(Lou Schmitz)

SFRB 6572 Rʀ-73
Class Rʀ-73 included series 5250-5599 and 6550-6599 all built at Topeka in 1962. With 135000 pounds capacity, 6572 was at Columbus, OH on the Pennsylvania RR December 4, 1962. Notice the different messages in the yellow left top placards on the three cars on this page. *(P.C. Winters)*

SFRE 29058 Rʀ-75
SFRE series cars differed from SFRBs with plug instead of sliding doors, also load dividers instead of DF bars to keep contents from shifting while in transit. Class Rʀ-75 comprised 100 cars, built by Santa Fe in 1962. 29058 was at Columbus on Dec. 26, 1962. *(P.C. Winters)*

SFRE 30013 Rʀ-74
Among 50 additional SFRE cars built by AT&SF car shops in 1962, 30013 was somewhere beneath the Pennsylvania Railroad's catenary when it was photographed. *(Bob's Photos)*

FLAT CARS

ATSF 90000 Fᴛ-X
Heavy-duty flat car 90000 was built in the AT&SF Topeka shop in 1953. With a load carrying capacity of 250,000 pounds, it had an overall length of 60 feet, 10 inches. It weighed 130,000 and the depressed center allowed extreme height loads to be transported. It was photographed at Columbus, in December 1962 with an electric transformer.　　　*(Paul C. Winters)*

ATSF 90003 Fᴛ-102
One of four heavy duty FS class flat cars built by Maxon Corp. in 1980 and among the last cars in age covered by our *ATSF Color Guide to Freight and Passenger Equipment*, 90003 had a 25 foot well and a load capacity of 238 tons, carried on eight axles.　　　*(D.P. Hobrook)*

ATSF 90016 Fᴛ-82
Santa Fe had four FD heavy duty flat cars of 81 foot length. With a 25 foot well for handling high and heavy loads, the transformer pictured on 90016 at Stowe, PA in January 1975 was safely handled by this flat built by Maxon Corp. in 1974.　　　*(C.T. Bossler)*

ATSF 90425 Fᴛ-73
Two special flat cars were constructed in company shops in 1970 to handle airplane bodies. 90424-90425 were classed as FBS with 94 foot 4 inches length and 70-ton capacity. 90425 was photographed in March 1971. *(Rail Data Services)*

ATSF 91971 Fᴛ-89
Santa Fe considered the 91930-91979 series as flat cars, however in the *Official Railway Equipment Register* they were assigned the A.A.R. GBSR gondola mechanical class. Equipped to handle coil steel they were built by Thrall Car Co. in 1976 and were 57 foot 8 inches over couplers with a capacity of 96 tons. *(Danny Sheardown, Ed Seay, Jr. collection)*

ATSF 90023 Fᴛ-95
This series included FM cars 90021-90023 built by Maxon Corp. in 1978 of 51 foot length, and a capacity of 250 tons for concentrated heavy lading. 90021 was photographed at Reading, PA in Novembr 1978. *(C.T. Bossler)*

ATSF 92956 Fт-6

▲ Built in company shops in 1955, 92956 was one of 100 early bulkhead flat cars of the LP A.A.R. Mechanical designation. Used primarily for wallboard or lumber lading, it was at St. Louis in September 1967.   *(P.C. Winters)*

ATSF 94283 Fт-19

▼ Fт-19 class were ordinary FM type flat cars of 53 foot 6 inches built in company shops in 1960 with 50 tons capacity. 94823 was at Norpaul, IL on the Indiana Harbor Belt Aug. 15, 1979 with a small caterpillar tractor.   *(D.F. Holbrook)*

ATSF 95843 Fт-32

▲ Two hundred 48 foot 6 inches bulkhead flat cars with *Super Shock-Control* cushion underframes were turned out of company shops in 1964, in series 95800-95999. Soon after completion, 95843 was earning revenues with a load of wallboard as it passed Summit in Cajon Pass.

(George Berisso collection)

ATSF 95932 Fт-32

▼ Another Fт-32 class bulkhead flat built in 1964, 95932 was handling the other commodity usually associated with this type of equipment when it was photographed at Aurora, IL on Burlington Northern in June 1980. *(D.F. Holbrook)*

ATSF 108477 Lᴳ-02

Serving the pine forests of East Texas and Louisiana, Santa Fe found use for some log flat cars of which 200 were built in series 108305-108504 in 1953 from gondolas originally constructed in 1923. 108477 was at Emporia in Oct. 1968, a long distance from the forests. One hundred flat cars were converted to log cars in 1956, the last examples built of the LG class.

(C.H. Humphreys)

 AUTO RACK CARS

ATSF Bi-Level

▲ AT&SF built an experimental auto rack car in 1960 at Topeka. By 1966, this new method of automobile transportation had increased the rail market share from 10% to over 50%. This Santa Fe bi-level had a black placard with gold Santa Fe logo when photographed in 1963 after passing over the Maumee River while on the NYC at Toledo, Ohio.

(E.J. Gulash)

ATSF 89396 Fᴛ-27

▼ The term *auto-veyor* never caught on as railroaders and shippers adamantly called them ''racks.'' 89396 was built at the AT&SF Topeka shops in 1963 and was photographed at Melvindale, MI in December 1964. It could accomodate 12 of the standard size autos of that period. *(E.J. Gulash)*

ATSF 88919 Fт-24

▲ This bi-level auto rack was built in Santa Fe shops in 1962 and was used to handle pick-up trucks when it was photographed in September 1971. 88919 was 87 foot 4 inches inside and 94 foot 5 inches over couplers and weighed 93,400 pounds light. It was photographed on Penn Central at Columbus, OH in September 1971. *(P.C. Winters)*

ATSF 88039 Fт-81

▼ One of 130 89 foot auto rack cars built in Santa Fe shops in 1973, 88039 was of the partially enclosed bi-level type for handling pick-up trucks. On Christmas Day 1976 it was at Council Bluffs, IA. *(Lou Schmitz)*

65

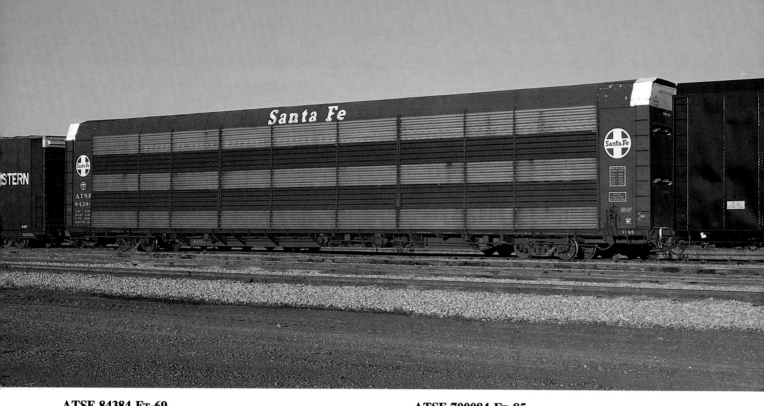

ATSF 84384 FT-69

▲ Feeling the only true protection was all-around enclosure, fully enclosed auto rack cars were on the scene by the 1970s. 84384 was one of 15 racks built by Pullman-Standard Car Manufacturing Co. in 1968. These tri-levels were 89 foot long and weighed about 46 tons empty. It was at Flat Rock, MI on July 6, 1974. *(Dave McKay)*

66

ATSF 700084 FT-85

▼ Whitehead and Kales delivered 160 of these *Safe Pak* full enclosed tri-level auto racks in 1974 that provided complete safety of the contents, except in derailments. These "monster" sized cars were 93 foot 8 inches in length over the couplers. 700084 was moving through Thorndale, PA on Penn Central on February 21, 1976, a short while before Conrail came into existence. *(C.T. Bossler)*

T.O.F.C. FLATS AND TRAILERS

ATSF 290543 Fт-26

▲ One of fifty 85 foot flats for trailer-on-flat-car service, built by General American Transportation Co. in 1962, 290513 was carrying two Santa Fe vans in November 1976. The American flag on SFTZ 290407B was applied for the 1976 Bi-Centennial celebration. SFTZ 202934 displayed the standard herald used on TOFC vans in 1976.

(Danny Sheardown, Ed Seay, Jr. collection)

ATSF 296530 Fт-87

▼ Bethlehem Co. supplied 296530, a 89 foot ''piggyback'' flat in 1974. 486 cars were in the Fт-87 class, series 296178-296663. It was a Streator, IL while carrying two trailers, one of which was a United Parcel Service van, to become one of the Santa Fe's best customers in the 1980s. *(D.P. Holbrook)*

ATSF 297220 Fт-37

68

▲ AT&SF Topeka car shop supplied forty-six 89 foot TOFC flat cars in 1965, in the series 297201-297325 which were not continuously numbered. Trailer SFVZ 700420 displays one of the later paint schemes used on ''piggyback equipment.''

(D.P. Holbrook)

SFTZ 205211

▼ This hi-cube trailer was painted with the standard herald then in use when it was photographed in September 1972 at Columbus, OH.

(P.C. Winters)

SFTU 500745

▲ Passing through Marion, OH in May 1979, SFTU 500745 was a container riding on a bogie. "U" indicated the box as a container, while "Z" was a trailer with its own wheels. Perhaps the *Piggy Back Service* slogan with red letters and emblem was the most attractive of various schemes applied to Santa Fe TOFC equipment.   *(P.C. Winters)*

ATSF 89989-B89989 Fт-16

▼ In the early era of "piggyback" service, company shops converted 133 1924 built 44 foot flat cars of class Fт-M to articulated flat cars for TOFC service, an example of which (ATSF 89989-B89989), was at Albuquerque in October 1974, 15 years after the conversion. On this trip one of the flats was a so-called "bare table."

A.T. Cruttendent, J.B. Moore collection)

69

ATSF 64724 Ga-144
A GS type gondola with drop bottoms built by Greenville Car Co. in 1965, 64724 had 70 tons capacity and was 53 feet 6 inches long. It was photographed at Reading, PA. January 30, 1971.
(C.T. Bossler)

ATSF 64949 Ga-102
The 39' drop-bottom fixed end gondolas of class Ga-102 were built by Santa Fe Topeka Shops in 1957. There were 175 cars of this class. The 64949 was handling a load of walnut logs from Kansas to Texas when it was photographed at Ardmore, OK on June 11, 1973.
(Earl Holloway)

ATSF 65476 Ga-82
Five-hundred 70-ton drop bottom GS gondolas were built for AT&SF in company shops in 1953. 65476 had a load of scrap iron and steel when its photo was taken in March 1975. Drop bottom gondolas were frequently used for various aggregates such as stone.
(M.J. Herson)

ATSF 70535 Ga-89

▲ AT&SF-owned ore cars, of which only 286 were on the roster Sept. 1, 1967 were normally assigned to copper ore loading in the Santa Rita, NM area on the New Mexico Division. 70535 was at Hurley, the crew terminal, on Nov. 25, 1978. It had been built by Pullman-Standard in 1953 and was 34 feet long. *(W.W. Childers)*

ATSF 70597 Ga-78

▼ An earlier ore car built by Pullman-Standard in 1950, 70597 was also at Hurley. Most of the ore cars had 95 tons capacity, including the two examples shown. *(D.P. Holbrook)*

ATSF 71192 Ga-13

A veteran GB class gondola 71192 was one of a series assigned to load sulphur at New Gulf, TX, and was built in March 1927. Over 30 years later in December 1968 it passed through Fort Worth on a southbound freight. There were 150 of these cars built by ACF with 70 tons capacity. *(K.B. King, W.W. Childers collection)*

ATSF 71859 Ga-26

A 1929 product of Standard Steel Car Co. 71859 had been modified for wood chip loading in East Texas, Louisiana, new Mexico and Arizona in 1961. It was at Gainesville, TX April 22, 1977. *(E. Holloway collection)*

ATSF 77927 Ga-123

The Ga-123 class encompassed only sixty 88 ton capacity hoppers built in the Topeka shop in 1963, when coal was an unimportant commodity on the AT&SF. 77927 was photographed in June 1977. *(Danny Sheardon, Ed Seay Jr. collection)*

ATSF 72073 Gᴀ-82
▲ Built in 1953 as a 54 foot gondola in company shops, 72073 was modified in 1964 for wood chip loading. In July 1969 it passed through Emporia, KS in front of the depot and Eastern Division office building.   *(C.H. Humpreys)*

ATSF 74000 Gᴀ-208
▼ Although built slightly beyond the time frame of this book, this experimental 111 ton all-aluminum hopper car that weighed only 40700 pounds is presented when it was new at Geneva, IL on January 3, 1981.   *(D.P. Holbrook)*

73

ATSF 76703 Ga-168

▲ These longitudinal (side) dump hopper cars of the HK mechanical designation were used primarily for company ballast traffic. Built by Greenville Car Co. in 1969, 76703 was at Abilene, KS Oct. 7, 1977. *(Dave McKay)*

74

ATSF 77338 Ga-64

▼ Another ballast hopper, 77338 was built in 1944 by ACF, one of an order for 400 "War Emergency" or "Victory" cars that required the approval of the War Production Board. It was photographed at Albuquerque Feb. 19, 1978 beyond the normal life of a freight car subjected to the hard service of a hopper car. *(B. Meunier, J.B. Moore collection)*

ATSF 78731 GA-56

▲ 78731 was hauling coal when it was at Columbus, OH in Feb. 1965. ATSF coal traffic was practically non-existent at that time. Two hundred 40 foot 8 inch, 70 ton capacity, three bay hoppers were included in an order from General American Transportation in 1958. *(P.C. Winters)*

ATSF 80875 GA-182

▼ A number of 70 ton covered hoppers were converted to open top hoppers for ballast service in 1973 at the Cleburne, TX car shop which handled open top car repairs. 80875 was at Arkansas City, KS April 2, 1975.

(D.P. Holbrook collection)

ATSF 81825 Gᴀ-176

▲ Low sulphur coal resumed moving from New Mexico mines in the late 1960s. 81825 was one of 200 triple hoppers built in 1971 by Greenville Car Co., of 100 tons capacity. It was hauling coke on one of its trips, at Cleveland, OH Jan. 16, 1972. *(Dave McKay)*

ATSF 86985 Gᴀ-156

▼ Gᴀ-156 class cars consisted of ten 53 feet 6 inches gondolas equipped with removable roofs for steel loading. 86985 was built by Greenville Car Co. in 1966 and was at Columbus, OH when photographed. *(P.C. Winters)*

ATSF 167340 GA-80
167340 was out of the Topeka shop in May 1973 and photographed a year later in good condition. GA-80 cars were 800 52 feet 6 inches gondolas with drop ends built by ACF in 1951 of 70 tons capacity, later increased to 77 tons.   *(P.C. Winters)*

ATSF 168402 GA-91
GA-91 class were 65 feet mill type gondolas of 70 tons capacity with steel floors of GM mechanical class. These 150 cars, including 168402 were built in the Topeka shop in 1955 and had friction bearings. Paul Winters lensed this car in Columbus, Ohio in April 1962.   *(P.C. Winters)*

ATSF 168998 GA-165
A more recent 65 feet mill type gondola with solid bottoms and fixed ends, 168998 was built by Santa Fe in 1968 and photographed June 19, 1976 outside Reading, PA on Conrail. Cubic capacity had been increased to 3287 feet, capacity to 90 tons, shock-control underframe installed and the now standard roller bearing journals applied.
(C.T. Bossler)

ATSF 169627 GA-61

▲ GA-61 class cars, series 169500-169699 were of composite wood and steel built by Pullman-Standard Car Manufacturing Company in war time 1943 to conserve steel. Of 70 tons capacity, 169627 was still carrying scrap iron at Allentown, PA in 1981. *(M.J. Herson)*

ATSF 176477 GA-53

▼ Built in Sept. 1941 by General American Transportation Co. the 400 GA-53 class gondolas were 53 feet over couplers, 50 tons capacity. 176477 was on the Santa Fe carfloat at Richmond, CA in 1959. *(K.B. King Jr., J.B. Moore collection)*

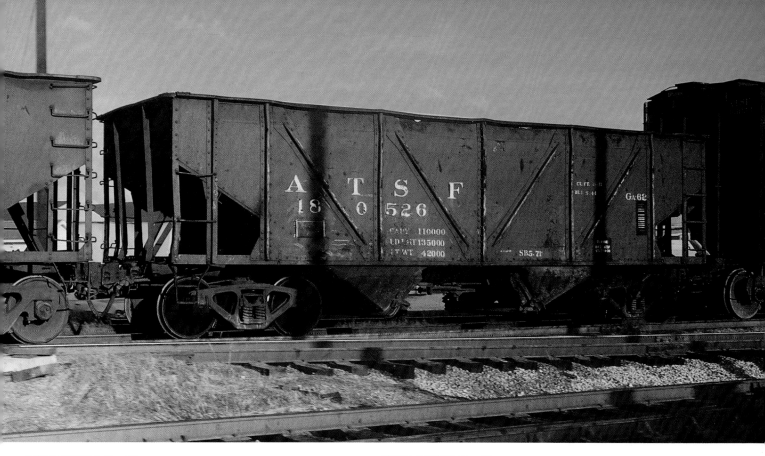

ATSF 180526 Gₐ-62

▲ 180526 was built as a composite wood and steel HM hopper in 1943 by General American Transportation Co. In 1958 steel side sheets were applied to these 200 cars. As improved, the 50 ton hopper was at Fort Worth in May 1973.

(K.B. King Jr., J.B. Moore collection)

ATSF 180607 Gₐ-60

▼ The Gₐ-60 class included series 180600-180799, two bay composite hoppers built by Pullman-Standard Car Manufacturing Co. in 1943. War veteran 180607 was still in service over three decades later when it was photographed at Ardmore, OK on Jan. 26, 1974.

(K.B. King, E. Holloway collection)

ATSF 180637 Gᴀ-60

▲ A view at Arkansas City, KS in April 1975 of 180637 another of the 1943 composite wood and steel hoppers, which had steel side sheets applied in 1958.

(D.P. Holbrook collection)

ATSF 186213 Gᴀ-193

▼ Air dump, side unloading cars, were almost always employed in company maintenance of way service. DIFCO built the thirty cars in the 186200-186229 with 50 cubic feet capacity in 1977 and 186213 was photographed at Abilene, KS on October 7 of that same year.

(Dave McKay)

TANK CARS

ATSF 98016 Tĸ-O

98016, after its 1953 construction by General American Transportation, spent most of its life hauling diesel locomotive fuel, but was later changed to transporting diesel lubricating oil. This series of cars were 47 feet 2 inches long and carried 16,000 gallons. *(W.W. Childers)*

ATSF 98274 Tĸ-M

Another car converted from its original purpose, 98274 built by General American Transportation in 1942, with a capacity of 15,933 gallons was built to haul steam locomotive fuel oil. AT&SF was the second largest operator of oil fired steam locomotives, next to Southern Pacific, and on one of its subsidiaries pioneered the successful use of oil in 1894. As photographed, it was assigned to hauling reclaimed diesel fuel oil.
(D. Akins,
W.W. Childers collection)

ATSF 98501 Tʀ-R

Tank car color indication bands circa 1970 indicated as follows: diesel fuel - grey, gasoline - red, car journal oil - yellow, water - green, diesel lube oil - yellow trimmed with grey, reclaimed lube oil - yellow trimmed in green, solvents - red trimmed in yellow. Tĸ-4 class tank cars were built by Richmond Car Co. in 1974 for diesel locomotive fuel hauling. With a capacity of 20,872 gallons, 98501 was a big, modern tank.

(W.W. Childers collection)

ATSF 100034 Tĸ-J

Another early tank car, 100034 was built by Pullman-Standard Car Manufacturing Co. in 1914 with a capacity for 10,580 gallons of locomotive fuel oil. It was still on the property on Dec. 22, 1972 at Ardmore, OK. *(E. Holloway)*

ATSF 100201R Tĸ-J

Santa Fe had a large "company fleet" of tank cars to haul locomotive fuel. For example, there were 600 tank cars in the Tĸ-J series when they were delivered in 1914. The last 29 were retired in 1981, at the age of 67 years. 100201 was at Ardmore, OK on Dec. 23, 1976.

(E. Holloway)

ATSF 100271 Tĸ-J

100271 had been built in August 1915 by Pullman-Standard Car Co. to handle 10,581 gallons of locomotive fuel oil from refineries to engine terminals. By that year, all locomotives burned oil in the territory west of Winslow, OK and south of Arkansas City, KS. It had been changed to company gasoline service when photographed at Abilene, KS in June 1971. *(W.W. Childers)*

ATSF 100591 Tк-K

▲ One of 500 12,000 gallon fuel oil tanks built by Pullman-Standard Car Co. in 1918 for locomotive fuel oil, 100591 was assigned to ''domestic water'' service at Nutt, NM on Nov. 25, 1978. *(W.W. Childers)*

ATSF 188373 Tк-D

▼ In domestic water service at Emporia, KS in March 1968, 188373 had been renumbered to company service series. This Tк-D car was built by ACF in April 1902, with a capacity of 9429 gallons of fuel oil. By 1902, all AT&SF steam power west of Williams, AZ had been converted to oil. *(C.H. Humphreys)*

ATSF 189076 Tᴋ-D

▲ Another former Tᴋ-D tank car, 189076 was in domestic water service and in MofW silver paint when photographed at Cleburne, TX.　　　　(D. Hale, W.W. Childers collection)

ATSF 202814 WK

▼ Domestic water tank car 202814 had been re-stencilled to WK (Work) class when it was photographed at Cleburne, TX May 16, 1976.　　　　(M.J. Herson)

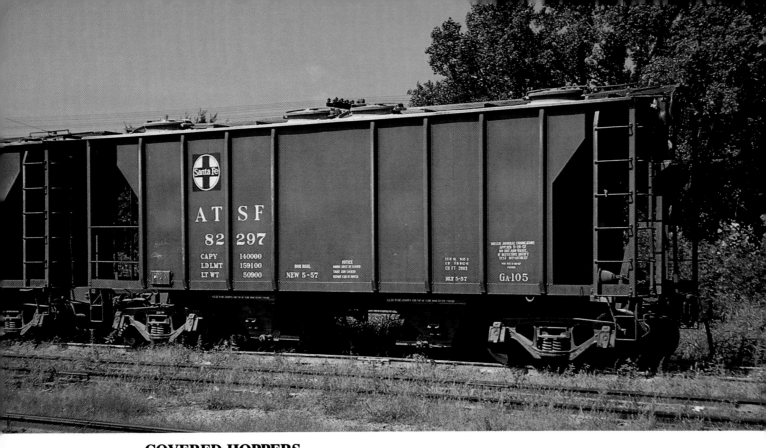

COVERED HOPPERS

ATSF 82297 GA-105
▲ Built by Pullman Standard Car Manufacturing Co. in 1957, 500 car GA-105 covered hoppers were of 2006 cubic feet capacity and 77 tons capacity. 82297 was still relatively clean from cement dust, the principal commodity handled, in August 1958. *(E.J. Gulash)*

ATSF 82697 GA-110
▼ The ACF-built 2006 cubic feet covered hoppers built in 1957 were painted gray and starting in 1959 certain classes of the LO mechanical designation were repainted grey. 82697 was photographed in May 1958 at an unknown location. *(E.J. Gulash)*

ATSF 87356 GА-101

▲ Five hundred 2003 cubic feet LO designated covered hoppers in the GA-101 class were built by Pullman Standard Car Manufacturing Co. in 1956 for the cement trade. 87356 was off line on the Wabash at Toledo, OH on May 5, 1962.

(E.J. Gulash)

ATSF 301411 GА-131

▼ A 2900 cubic feet 100 ton capacity car, 301411 was one of 200 built in 1962 by ACF, equipped with roller bearings. These small cubic capacity round type LO cars were used primarily to haul potash or fertilizers and other dense bulk commodities. It was photographed at Emporia, KS in October 1974.

(C.H. Humphreys)

ATSF 101458 GA-178

▲ In 1972, ACF built 100 covered hoppers of a unique design to transport bulk potatoes from California. With 90 tons capacity and 4600 cubic feet area, they were equipped with mechanical refrigeration to keep the temperatures at about 40 degrees F. The noble experiment of the *Conditionaire* cars was a failure account bruising to the "spuds" when they were loaded and unloaded. The cars were transferred to grain and other bulk commodity service, with of course the refrigeration units removed. A still fresh 101458 was at Pacific Junction, Iowa on July 2, 1972, in its as-intended service.

(Lou Schmitz)

ATSF 101434 GA-178

▼ A much more tarnished 101434 was at Ardmore, OK on November 8, 1980 not in need of its refrigeration capability. The dark "reefer" orange and outside insulation weathered very badly.

(Earl Holloway)

ATSF 300734 G<small>A</small>-116

▲ Built in May 1959, G<small>A</small>-116 300734 was at Fort Worth, TX in December 1962. The grey colored hopper had ten round roof hatches. *(K.B. King, E. Holloway collection)*

ATSF 301085 G<small>A</small>-119

▼ A 70-ton LO covered hopper also built in 1959 by Pullman-Standard Car Manufacturing, 301085 was a fore runner of a fleet of covered hoppers that would eventually replace box cars for grain loading. It was at Sioux City, IA Sept. 27, 1977 with original paint. *(D.P. Holbrook)*

ATSF 301942 Gₐ-135
One of an order for 312 LO covered hoppers equipped with trough hatches, 301942 held only 4000 cubic feet capacity, but soon demonstrated its superiority over XM box cars for grain loading. Built in 1963, it was almost new when photographed at Columbus, OH that September. *(P.C. Winters)*

ATSF 305101 Gₐ-150
Among the larger LO covered hoppers, series 305100-305119 of the GA-150 class were built by ACF in 1965 with a cubic capacity of 4650 feet. Somewhat shabby, it had not been repainted when photographed at Peoria, IL on Nov. 11, 1980.

(D.P. Holbrook)

ATSF 305874 Gₐ-151
The new "work horse" of the freight car fleet were the standard 4427 cubic feet LO covered hoppers as represented by 305874 one of 1500 built in 1965 by Pullman-Standard Car Manufacturing Co. *(P.C. Winters)*

ATSF 310000 GA-93

90

▲ The 50 GA-93 class LO type cars were built in 1955 by General American Transportation and assigned to the cement trade. 310000 had been recently shopped when it moved through Emporia, KS in January 1973. *(C.H. Humphreys)*

ATSF 310137 GA-124

▼ A later 2600 cubic feet LO hopper, 310137 came from General American Transportation in 1961 as part of an order for 52 cars. It was in Muhlenberg Township, Pennsylvania on May 15, 1973. *(C.T. Bossler)*

ATSF 310676 Gₐ-205

One of 50 large 4180 cubic feet air-slide covered hoppers built for bulk flour loading 310676 came from General American Transportation in 1979. By the early 1960s most bakery flour was shipped in covered hoppers, which were unloaded with air pressure. *(M. Spoor)*

ATSF 311891 Gₐ-183

Upon delivery of 1000 additional 4427 cubic feet covered hoppers in 1971, AT&SF rostered a total of 6900 of these covered hoppers, which were kept very busy handling wheat to the Gulf ports beginning with the heavy moves of 1972-1973 to the Soviet Union. 311891 was new at Cleveland, OH on January 12, 1971. *(Dave McKay)*

ATSF 313831 Gₐ-183

Cubic capacity was increased to 4600 feet with the delivery of 1000 ACF covered hoppers in the series 313800-314799 built in 1973. Weight capacity remained at 100 tons. The photo was taken on November 28, 1974. *(Ed Seay Jr.)*

REFRIGERATOR CARS

SFRD 5011 RR-26

▲ The 10 RR-26 class cars were modified in 1939 from RR-10 cars of 1931. They were equipped with 7 inches of insulation to protect frozen food shipments. These were 50 foot 8 inches overall in length, with 90,000 pounds capacity and equipped with ice bunkers that could hold 14,500 pounds of crushed ice. 5011 was still on the car roster at Corwith Yard in Chicago Sept. 10, 1972, but mechanical refrigerators had long ago took over frozen food lading.   *(Dave McKay)*

SFRD 10291 RR-46

▼ Built in 1923 as wood sheathed, USRA design cars, class RR-46 including 10293 were all extensively rebuilt at West Wichita car shop in 1948-1949. They were equipped with ventilating fans that worked when the car was in motion. There were 199 of this class in Jan. 1971 and the last one was retired in 1977.   *(George Berisso collection)*

92

SFRD 10334 Rr-46

▲ Some vegetable shippers preferred ''icers'' for leafy vegetables. 10334 had the large Santa Fe billboard emblem when it was photographed at Chicago Sept. 10, 1972, a year when only 100 of the Rr-46 class were still active of the 1000 rebuilt in 1948-1949. At this late date it was used only for ''dry'' freight as the icing stations were all closed by 1972.

(Dave McKay)

SFRD 11634 Rr-49

▼ ''RD'' 11634 was awaiting light weighing after being shopped at San Bernardino in October 1958. The *route of the San Francisco Chief* slogan was started in 1954 with the inauguration of the last new Santa Fe streamliner. Stage icing permitted a section of the bunkers to be separated from the lower part when it was desired not to fill the bunkers for perishable shipments not needing cooler temperature. It had been rebuilt in 1950 from a wood sheathed Rr-5 built in 1927.

(K.B. King Jr., Ed Seay, Jr. collection)

93

SFRD 13241 Rr-51

 ▲ Originally a wood body car built by Pullman-Standard in 1928, 13241 had been rebuilt to an all-steel car in 1952 and was still on the roster in July 1969, one of 158. Santa Fe officially painted its refrigerator cars in "reefer yellow No. 26" color.

(Rail Data Services)

SFRD 16595 Rr-23

▼ This reefer was on its way east through Emporia, KS in September 1972 to be scrapped. 16595 had been built as class Rr-W in 1919, rebuilt in 1937 and modernized in 1955, with this number. It was one of only two Rr-23 cars left at this date.

(C.H. Humphreys)

SFRD 19626 Rʀ-33

▲ Originally a United States Railway Administration design "reefer," 19626 was rebuilt at West Wichita in 1941, as part of a 500 car rebuilding program to change them to all-steel sheathed cars with ventilating fans. It was a survivor when it passed through Emporia in September 1972. Only 35 Rʀ-33 cars were left by January 1, 1973, all using 50 year old underframes. *(C.H. Humpreys)*

SFRD 20699 Rʀ-29

▼ SFRD 20699 was among familiar company as it passed through Columbus, OH on a Pennsylvania freight train December 26, 1962. The RS type ice "reefers" still handled a large portion of the West Coast fruit and vegetable crops. Rʀ-29s were new all-steel cars built by General American Transportation in 1940, in the series 20500-20792. These cars were equipped with Duryea Cushion underframes. *(P.C. Winters)*

95

SFRD 8148 Rʀ-40

 ▲ *Ship and Travel SANTA FE - all the way* was the slogan in black script on this reefer. This slogan was adopted commencing in April 1947. Yellow-orange sides and black roof and ends were SFRD trademark colors on their reefer fleet. 8148 was at Council Bluffs, IA on June 13, 1965.

(Lou Schmitz)

SFRD 33110 Rʀ-23

▼ In 1937, Santa Fe commenced to rebuild refrigerator cars after a depression hiatus and 1500 steel-sheathed cars were turned out in company shops from USRA design wood-sheathed reefers. 33110 was still around in January 1971, but only one other car in this class remained. *(Dave McKay)*

SFRD 2103 Rʀ-56

▲ Early mechanical refrigerators were designated SFRDs, but later changed to SFRP about 1961. One of 175 cars built at West Wichita in 1955, they followed the first 29 mechanicals built in 1953. This ''reefer'' was still in a nice coat of the original orange body paint when photographed with dark blue doors and white ''MTC'' slogan. Note the black roof but orange ends at Columbus, OH December 4, 1962.

(P.C. Winters)

SFRC 1218 Rʀ-69

▼ Class Rʀ-69 included 500 RPL cars built by General American Car during 1961. 1218 was still in its large logo original livery at Columbus, OH Nov. 21, 1962. These RPLs were 56 foot, 6 inches overall length, with a load capacity of 128,000 pounds and cubic capacity of 3085 feet. Trane-type mechanical refrigeration was used.

(P.C. Winters)

97

SFRC 50305 Rʀ-87

▲ In 1965, 395 mechanical refrigerators were built in Santa Fe shops, including the 50300-50694 series. These were large cars of 60 foot, 6 inches length over the couplers and a load capacity of 124,000 pounds and cubic capacity of 4074 feet. All-purpose temperature range allowed their use for all perishable commodities. *(E.J. Gulash)*

SFRC 50730 Rʀ-89

▼ Built in 1966, 50730 was in practically new condition as it was photographed in October of that year. A total of 700 RPL Rʀ-89 cars were constructed by Santa Fe in 1966. Apparently the introduction of mechanical reefers started a trend toward a darker orange for Santa Fe reefers as is apparent in this photo. *(Bob Wilt)*

SFRP 2154 RR-56
In June 1973 in Toronto, Ontario, 2154 was still in its small logo circa 1955 paint and lettering. SFRP cars had no load restraining equipment, but otherwise were similar to the 56 foot, one inch long SFRCs. *(Dave McKay)*

SFRP 2279 RR-56
SFRP 2279 was originally built in 1955 with SFRD reporting marks to haul frozen foods (especially strawberries). By the time it was photographed in March 1964, it had been repainted with the nine foot diameter Santa Fe herald. SFRD became the reporting marks only for ice activated "reefers."
(P.C. Winters)

SFRP 2194 RR-60
1955 saw the first big acquisition of "MTC" (mechanical temperature control) refrigerators with 231 built in the West Wichita shop. 2194, shown at South Omaha, NE July 10, 1973 was one of 50 RR-60 class cars added in 1958. *(Lou Schmitz)*

STOCK CARS

100 **ATSF 23003 Sᴋ-4**

▲ In 1950 company shops built 200 double deck SC type stock cars from automobile cars of 1923 construction. 23003 was the fourth car turned out and was one of 65 left when it was photographed at Emporia in August 1969. AT&SF double-deck cars were used primarily to handle sheep and hogs, however by 1950 most hogs were trucked to market. Note the reefer yellow doors on double-deck cars. *(C.H. Humphreys)*

ATSF 26668 Sᴋ-2

▼ Single deck 26668, of the SM mechanical designation was built in 1942 from a retired box car built in 1923. The last new stock cars, 300, were purchased from builders in 1930. Thereafter other cars were rebuilt to stock cars when replacements were necessary. *(Steve Evarts)*

ATSF 27026 Sᴋ-3

▲ A view of 27026 after it was condemned at Albuquerque in March 1974. It was one of the last survivors of 1120 class Sᴋ-3 single deck stock cars rebuilt from box cars Bx-8, 9, 10 during 1947. It was of the standard 40 foot, 6 inches length, as were all AT&SF stock cars built commencing in 1906. Long after all 36 foot cars were retired, tariffs permitted shippers to order them and the railroad had to protect the lower minimum weight, saving the shipper freight charges. The original box car ends were used in the reconstruction of these cars.   *(A.T. Cruttendent, J.B. Moore collection)*

ATSF 27905 Sᴋ-3

▼ Another Sᴋ-3 class stock car, 27905 was still on the roster when photographed in March 1973. AT&SF had cancelled its livestock tariffs in 1971 however, and it would soon move to the scrappers. This veteran survived because it was used as a "reach car" by crews switching the Navajo oil refinery at Artesia, NM where diesels were outlawed.   *(Steve Evarts)*

ATSF 28415 Sk-5
The 1300 Sk-5 class single deck stock cars built in 1952 from retired box cars were the last added to the AT&SF fleet. In that year, 80,324 cars of livestock were handled and at the end of the year, 7402 stock cars were on the car rosters. 28415 was still on the property in January 1973, but would soon be sold for scrap.
(Steve Evarts)

ATSF 28487 Sk-5
Fortunately railfan photographers took some photos before all AT&SF stock cars disappeared. 28487 was photographed in September 1973, one of a handful of Sk-5 cars that were on the property. It was at La Junta, CO and highly modified for its new life in sugar beet service.
(Rail Data Services)

ATSF 50700 Sk-R
Five-hundred single deck cars were built by Pennsylvania Car Co. at Argentine, KS in 1928 of the Sk-R class, series 50500-50999. At Waynoka, OK August 5, 1973, 50700 was one of the last still on the property. Some cattle was still being moved in 88 foot private-owned cars, but AT&SF was no longer furnishing its cars and had retired its stock pens.
(E. Holloway collection)

ATSF 60272 Sk-T

▲ Pennsylvania Car Co. was the builder of 500 SM stock cars in 1929 of series 60002-60501. Veteran 60272 was still in service showing excellent car end details when it moved through Emporia, KS in August 1969. 167 of these cars made it to 1970. *(C.H. Humphreys)*

ATSF 69211 Sk-Z

▼ In 1941, company shops produced 300 double-deck stock cars from 40 foot automobile cars built in 1923. Numbered 69101-69300 they were the last to use an all alpha classification. The next group of stock cars, built in 1942 were Sk-2, no Sk-1 being assigned. 69211 was photographed in December 1967. *(Rail Data Services)*

103

ATSF 60393 Sᴋ-T

During the grain car shortages of the 1960s, some stock cars were fitted with plywood interiors and used to handle grain. 60393 is an example of this temporary conversion until more covered grain hoppers were received. Due to their low capacity the cars were unpopular with shippers, but were better than storing grain on the ground. This Sᴋ-T was at Hutchinson, KS with a grain elevator in the background in March 1968.

(C.H. Humphreys)

CABOOSES

ATSF 1785R

▲ Caboose 1785R was in a "plain Jane" mineral brown motif at San Angelo, TX in August 1961. Built by ACF in 1929, it was rebuilt to a 999000 series car in 1967. R indicated radio-equipped. *(K.B. King Jr., J.B. Moore collection)*

ATSF 1579

▼ A safety slogan placard relieves the all-mineral brown of caboose 1579 at Corwith in September 1964. 1579 was among the first 100 all-steel cabooses built by ACF in 1927.

(Dave McKay)

ATSF 1910R

▲ 1910R was among 104 all-steel cabooses built by ACF in 1930 and was at La Junta, CO on October 12, 1969.

(Dave McKay)

ATSF 999000 Cᴇ-1

▼ The first of the rebuilt pool cabooses of 1966 was the 999000. An agreement with the conductors and trainmen was effective in 1967 that eliminated the assignment of a caboose to each crew, therefore avoiding changes at division points. Train crews appreciated the cushion underframes that reduced shocks from slack action. Since this feature was utilized, the cars were painted Indian red instead of mineral brown.

(Bob Wilt)

ATSF 999700 CE-8
Santa Fe ordered 50 wide-vision cupola cabooses from International Car in 1977, of which 999700 was the first. It was photographed on one of its early trips in February 1978.   *(Danny Sheardown, Ed Seay Jr. collection)*

ATSF 999540 CE-6
The first of the wide-vision cupola cars were five built in 1974 by International Car. This series was 999538-999542. The 999540 was at Corwith Yard in Chicago on March 20, 1977.
          (Dave McKay)

ATSF 999471 CE-2
Cabooses were rebuilt at West Wichita and San Bernardino. 999471, was a San Bernardino shop product of 1968 but its paint was still in good condition at Prescott, AZ on October 13, 1976.   *(Dave McKay)*

ATSF 999716 CE-8
Another International Car wide-cupola car, 999716 was at Tulsa, OK in May 1980, still in pristine condition. *(Dave McKay)*

ATSF 999214 CE-1
The 999214 at La Junta, CO on October 12, 1968 was recently rebuilt at the West Wichita car shop from a 1750-1874 caboose built by ACF in 1930.
(Dave McKay)

ATSF 999651 CE-7
The 999651 was one of twelve modified CE-7 class cabooses for local train service in 1974. It was at Flagstaff, AZ on October 12, 1976. *(Dave McKay)*

ATSF 1708 Cw-1

▲ Caboose 1708 was built by ACF in 1928. By June 1974, it had its cupola removed and was assigned to transfer service.

(Danny Sheardown, Ed Seay, Jr. collection)

ATSF 2243 Ce-1

▼ 2243, was one of 100 "way cars" built in company shops in 1948, in the 2201-2300 series. At Saginaw, TX June 27, 1976, it still was painted mineral brown.

(E. Holloway collection)

ATSF 2224

▲ 2224, another Ce-5 class caboose at Saginaw June 27, 1976 had an experimental paint scheme that was used on only a few cars. Red and white were the only colors used.

(E. Holloway collection)

ATSF 999607 Ce-3

▼ A much more colorful 999607, a class Ce-3 caboose was assigned to local freight service. A yellow cupola indicated use in local service. A white cupola indicated use in run-through Santa Fe/Penn Central service. 999607 was at Gaines-ville, TX Aug. 24, 1975.

(E. Holloway)

DROVERS CARS

ATSF D-936

Drovers cars were built to accommodate livestock caretakers, who rode them instead of the caboose. D-936 on October 12, 1968 at La Junta had been modified in 1942 for branch line local freight service and numbered 2314. In 1959 it was renumbered back to D-936 but its function did not change.

(Dave McKay)

ATSF 932-D

932-D, one of nine drovers cars left in 1967, was built by ACF in 1931 and was never modified for branch line freight service. The photograph taken at La Junta. *(C.H. Humphreys)*

ATSF 2320

Built in 1935 the 49'5'' 2320 was rebuilt to rider coach 2599 in 1942. It was further modified to combination baggage-coach 2320 on December 17, 1946.

(K.B. King, E. Holloway collection)

 ## WORK EQUIPMENT

ATSF 188248 Iᴇ-X

▲ A retired box car fitted with refrigerator car doors was used to make ice car 188248 which was found near the Glendale, AZ icing plant in 1960. The Iᴇ-X class designation came from IcE car, from boX car.

(K.B. King, W.W. Childers collection)

ATSF 188457 Wʀ-R

▼ This ice car for track work gangs was converted from a 1937 built Rʀ-21 class refrigerator (hence class Wʀ-R) built by General American Transportation Co. 188457 was in its newly applied aluminum paint at Gallup, NM in 1960.

(K.B. King, W.W. Childers collection)

ATSF 189114

▲ Ex-steam locomotive tender 1354 which came to the AT&SF in 1912 behind that 4-6-2, and later served with 2-6-2 1857 was placed behind Emporia, KS derrick 199784 in 1954. It was still on hand at Cleburne, TX in 1971.

(K.B. King, W.W. Childers collection)

ATSF 190589 Wᴋ-O

▼ Assigned to signal gang unit no. 11 at Richmond, CA during the 1950's, ATSF 190589 was used to transport track motor cars and was a former emergency road wheel and tool car. (K.B. King, W.W. Childers collection)

ATSF 190985 Wᴛ-G

Incorrectly stencilled WX, 190985 was a rail and tie car with boxes on each end containing track material such as spikes, tie plates, joint bars, etc. It was assigned to the Cleburne derrick and was at Fort Worth in 1968.

(K.B. King, W.W. Childers collection)

ATSF 191305 Wᴛ-G

Another combination rail-tie car, 191305 was assigned to the San Angelo derrick outfit in 1956.

(K.B. King,
W.W. Childers collection)

ATSF 191403 Wᴋ

Boom car 191403 used with a wrecking derrick was a former baggage car with superstructure removed. It was assigned to derrick 199797 which was located at Temple, TX in 1973.

(M. Spoor collection)

ATSF 194450 Wk

Business car 10, built by Pullman Co. in 1928, was converted to foreman's car 194450 in 1967. At Needles, CA June 1, 1973, this foreman enjoyed more room and luxury than most of his co-workers. *(J. McMillan, W.W. Childers collection)*

ATSF 195365 Wk

Foreman Car 195365 was rebuilt from box car Bx-14 226461 at Albuquerque in 1945. It was photographed at San Angelo in 1960. A propane gas heater has been recently installed, but there was no air-conditioning.

(K.B. King, W.W. Childers collection)

ATSF 198741 Wk

Kitchen-dining car 198741 was former baggage-mail car 2084 before it was rebuilt for work service in March 1952 and assigned to the Cleburne derrick outfit. It was at Fort Worth returning from a derailment in 1970. *(K.B. King, W.W. Childers collection)*

ATSF 199202 Wᴋ

Weed spray car 199202 was converted from baggage car 2050 at Cleburne in 1964. The chemical tank cars were entrained next to the sprayer. The unit was painted in the blue-and-yellow freight diesel colors. It is shown at Ardmore, OK on April 21, 1976.

(Both, Earl Holloway)

ATSF 199205 Wₖ

This converted Railway Post Office car, 199205 was classified as a brush sprayer. The conversion work was done in January 1968. Photographed in January 1972, it was not retired until April 1985. *(K.B. King, W.W. Childers collection)*

ATSF 199229

Ballast spreader 199229 was built in April 1923 as the 199553. It was on hand at Summit, CA on Cajon Pass on Jan. 29, 1956. Retirement would come in August 1963.

(E. Waddingham, Bob Trennert collection)

ATSF 199232

Jordan spreader 199232 was the second machine to carry that number. This modern spreader was built in Oct. 1961 and photographed at Williams, AZ on Nov. 11, 1963. *(Bob Trennert)*

ATSF 199296 Wx-30

Ballast cleaner 199296 was mounted on a gondola car body and was designated as an "in-track conveyor." It was at Albuquerque in Sept. 1972.

*(K.B. King,
W.W. Childers collection)*

ATSF 199361

One of the most interesting pieces of non-revenue equipment is rotary snow plow 199361 which was built at Topeka in Nov. 1959 by mating the tender from retired 4-8-4 3769 with the rotor and shroud from old 1892 built Cooke rotary 199398. It was powered by electricity from a diesel-electric unit. As many additional units as necessary push the plow. As this is written in 1995 it is still held at Topeka between its infrequent assignments. *(Dave McKay)*

ATSF 199368

Wedge plow 199368 was at the Topeka shop on July 19, 1964, after its retirement in 1960. This type of plow was stationed at branch line terminals in Illinois, Missouri, Kansas, Colorado, Oklahoma, Texas Panhandle and New Mexico and were quite often shoved by the locomotive on the regular local or mixed train when cuts filled with snow. Note the "load" of stone ballast and weed garden.

(W.W. Childers)

ATSF 199390

▲ Russell snow plow 199390 was built in 1925 and still on-hand at Williams, AZ on Nov. 9, 1963 if needed for snow plowing on the Arizona divide. *(Bob Trennert)*

ATSF 199598 Wᴛ-J

▼ Rider car 199598 used with tie unloader 453 was at King-man, AZ on Jan. 14, 1972. Most tie replacement was done in the spring-summer-early fall and the equipment was stored the balance of the year.

(J. McMillan, W.W. Childers collection)

119

ATSF 199754

A 115 ton derrick, built for the Kansas City, Mexico & Orient Railway, which AT&SF bought in 1928, 199754 was built in 1915 by Bucyrus. It was still on hand at Cleburne in Nov. 1970.   *(K.B. King, W.W. Childers collection)*

ATSF 199793

Among the largest derricks used on the System, blue and yellow 199793 was a 200 ton capacity Industrial Works product of 1927, originally steam powered, but later converted to diesel-electric. It was assigned to Newton, KS for many years, where it was photographed. Increasing use of contractors with off-track equipment to clear wrecks, resulted in retirement of this "big hook" in June 1988.

(Dave McKay)

ATSF 199797

Zebra-striped 200 ton capacity derrick, 199797, built by Industrial Brownhoist in 1931, spent its life at Temple and Cleburne, and was at Temple in December 1973. It was also retired in June 1988.   *(R.L. Roark Jr., M. Spoor collection)*

ATSF 203211 Wк-48
Living or bunk car 203211 was a latter day box car conversion. It was at East Galesburg, IL in June 1977.   *(W.W. Childers)*

ATSF 205991 Wк-50
A former horse car, equipped with end doors, 205991 was a "Roadway Machine Parts Car" used for handling parts for the many track machines on the railroad. It was photographed at Temple, TX on Apr. 20, 1979.
(W.W. Childers)

ATSF 206883 Wт-16
A track panel car, 206883 was sent to derailments on a work train to replace damaged trackage. It was at Brownwood, TX in 1980.
(K.B. King, W.W. Childers collection)

ATSF 206982

 ▲ The 206982 was the boom car for the Barstow derrick outfit, and was photographed May 21, 1980, with derrick 199791 which had a 200 ton boom capacity. *(Dave McKay)*

ATSF 206997

▼ A converted baggage car, Santa Fe 206997 was also at Barstow on May 21, 1980. Stencilled "Barstow Tool Car" under its large black logo, the car wore the aluminum paint Santa Fe used on MofW equipment. *(Dave McKay)*

ATSF 199966

▲ Diesel locomotive traction motors were overhauled at San Bernardino, requiring traction motor carrier cars such as the 1999666. It was at Cleburne on January 11, 1976, ready to be utilized to haul motors out to the other heavy repair shops at Cleburne and Argentine, KS. *(Matt Herson collection)*

ATSF 205228 Wx-36

▼ After many years of handling revenue tonnage of numerous classifications, renumbered 205228 had been assigned to company material service and painted in the latter day grey paint for non-revenue cars. 205228 was photo'ed at Summit, CA. *(Matt Herson collection)*

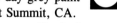

ATSF Work Train
▲ Santa Fe recycled its equipment as newer additions arrived. The cars of this typical work train at "Chilli" in August 1963 were once top of the line revenue-producing rolling stock. The end car is a tool car, while the heavyweights are bunk cars. *(Russ Porter)*

ATSF 198953
▼ Many heavyweight passenger cars found additional employment as MofW equipment after being retired from revenue use. 198953 was a former Railway Post Office Car on the work train at Chillicothe, IL in August 1963.
(Russ Porter)

SPECIAL EQUIPMENT

ATSF 30

▲ Santa Fe built 30 remote control units, termed RCEs, from F3 and F7 booster units from 1967 to 1969. The 30 at Barstow in Sept. 1974 is an example. Their operation, after some early service in Argentine-Barstow symbol train service, was subsequently limited to a few heavy unit trains, carrying sulphur in tank cars, grain and potash.   *(Ed Seay Jr.)*

ATSF 18

▼ A much dirtier 18 was photographed in Texas in February 1979. The B units were stripped of their prime movers and fuel tanks and carried only the radio equipment, but their former locomotive status permitted their use in this heavy duty service.   *(Danny Sheardown, Ed Seay Jr. collection)*

ATSF 1-5

▲ Santa Fe built five steam generator cars in 1943 from 1337-class 4-6-2 type locomotive tenders for use with the 5400 h.p. FT freight diesels when it was necessary to press them into troop train service on the Coast Lines. Originally numbered 9000-9004 they were re-numbered to 1-5 in August 1967, shortly before two were retired that September, with the balance following in June 1968. All five were still on hand at White Eagle, OK on Aug. 11, 1969.   *(L.E. Stagner)*

ATSF 4

▼ Steam generator car 4 was on hand at Cleburne, TX in Dec. 1967 before its June 1968 retirement. They were equipped with two Vapor-Clarkson steam generators with a capacity of 2250 pounds per hour and had a capacity of 8700 gallons water and 1075 gallons of fuel oil. Entry to the generators was accomplished through the doorway between the windows. The four port holes are reminiscent of the FTs they often served behind in the 1940s.   *(Rail Data Services)*

Dynamometer Car 29

▲ The 29 was built by the Santa Fe in 1910 as its first dynamometer car to road test the drawbar pull of steam power, from 2-6-2s to 2-10-4s. Its last road test with a steam locomotive was in 1948 when newly equipped poppet valve 4-8-4 No. 3752 was given extensive tests in both freight and passenger service. It also was used to test the early diesels, including EMD FT demonstrator 103 in 1940. It was seldom used after 1948, and was photographed at Topeka shops on January 2, 1954 but not retired until February 1963. 29's successor was test car 5015, put in service in September 1961 and better adapted to test diesel locomotives. *(Lou Schmitz)*

Rail Detector Car

▼ "Tow" unit 9166 provided the power to propel Rail Detector Car 3, which were at Barstow on February 4, 1977. Built as the third AT&SF rail detector car in 1942, the original gasoline engine in 9166 was replaced with a 165 h.p. diesel in 1967. The residual magnetic detector system used on the 3 was upgraded to the ultrasonic method in the early 1970s. In the early 1980s AT&SF hired the Sperry rail detectors and retired their own equipment.

(Joseph R. Quinn, David Hickcox collection)

127

 *This turquoise shield graced the wall of the private dining room in the 500-505 series dome-lounge cars built by Pullman in 1950 and assigned to the SUPER CHIEF. Amtrak had not removed the shields when this photo was made in 1973, although the service provided did not include the use by private parties of the Turquoise Room. What remained was memories of first class service on one of America's most famous and luxurious trains—the SUPER CHIEF. Just a small example of the first class equipment utilized on the Atchison Topeka & Santa Fe!*

(Dave McKay photo)